Cells For Kids (Science Book For Children)

By Nishi Singh, PhD

Table of contents

1. What is a cell?

A cell is a **basic building block** of all organisms (living things). They can also be called the **smallest unit of all things living**.

Living things are organisms that respond to changes in the environment. All living things are those that need nutrition, move and grow, reproduce, breathe and respond to the external environment.

Humans, plants and animals are living things. The cells in living organisms form the basic structural, functional and biological unit.

A cell contains the molecules of life and is the central unit of biological organization. Each cell type has a different job to do.

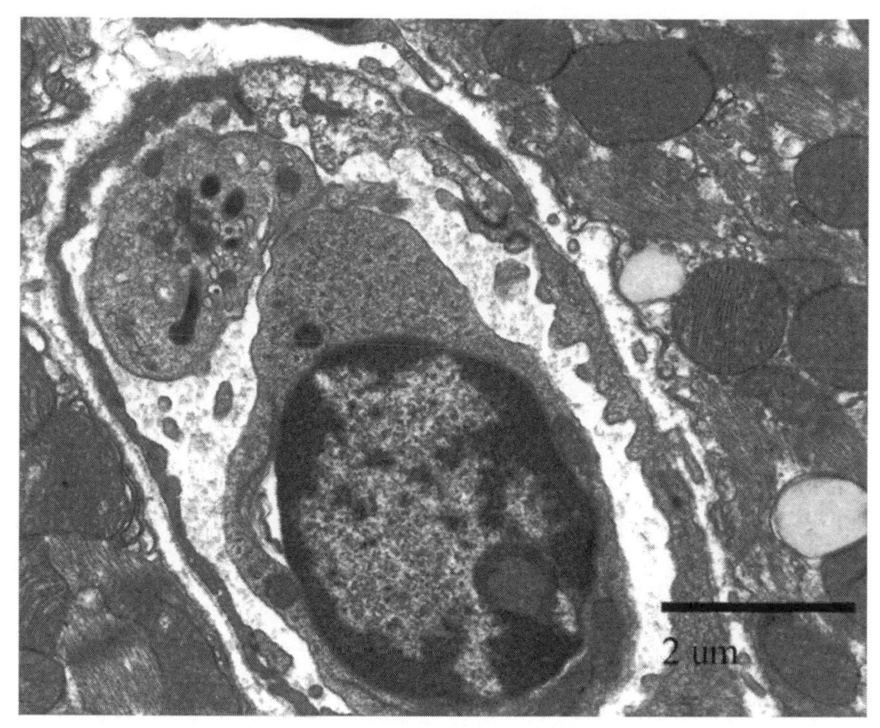

This is a cell from a mouse's heart muscle

Questions:

1. What is a cell?

2. What are living things?

2. Who discovered the cell?

Robert Hooke discovered the cell on 18th of July in the year 1635. He was often very sick so he spent a lot of his time schooling at home. As a child he was very curious and wanted to know how the body worked.

As an adult, he worked as an assistant to the English scientist Robert Boyle. He spent a lot of time looking under the "**microscope**". While he was looking at a "**cork**" under the microscope, he noticed that the cork was made of compartments which he called "**cells**".

He drew what he saw and called them "cells" to describe them because it looked like empty rooms where monks used to live. He wrote a book called

“**Micrographia**” in September 1665 about his observations.

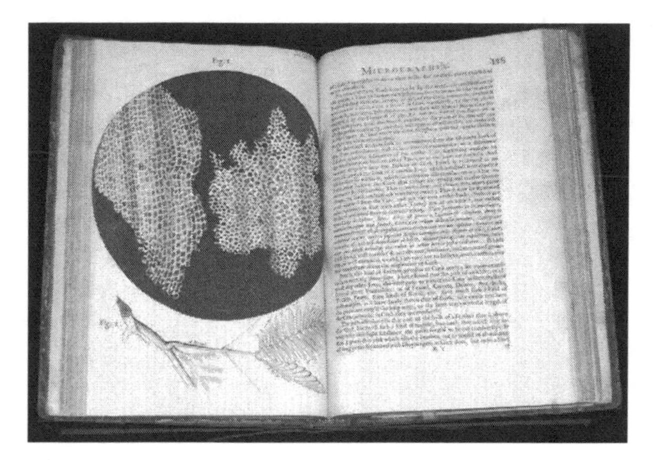

Questions:

1. Who discovered the cells?

2. What was the name of the instrument he used?

3. Why did he call what he saw under the microscope as “cells”?

4 What was the name of the book he wrote?

3. What are cells made of?

The cell is made of a cell membrane that holds the cell in place and prevents it from falling apart. The cell has a nucleus that controls the cell's activities. It also has a jelly-like material called the "**cytoplasm**". It has other components called "**organelles**" which you will learn later in the book.

Cells are mostly made of water. About **two thirds** (2/3) of the cell consist of water. This means that 2/3 of your whole body is also made of water. The rest of the cells are made of protein, fats, carbohydrates, DNA and other molecules.

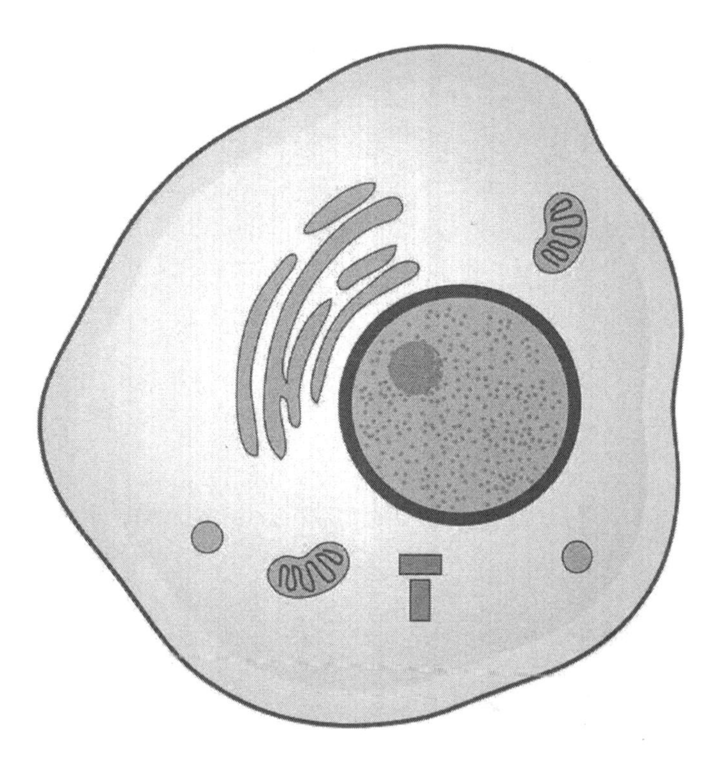

Questions:

1. What are cells made of?

2. What portion of the cell is water?

3. What are cell components called?

4. Which is the jelly-like substance?

4. Why cells are mostly made of water?

A cell is water (70%) while the rest of the cell components (30%) contain structural and functional molecules.

In order for a cell to retain their shape, they need to be in water and also be surrounded by water.

The chemical process inside the cell takes place in solution. So cells must have a watery environment.

The components of a cell are molecules that are mostly dissolved in water. The watery environment is full of cellular machinery and structural elements.

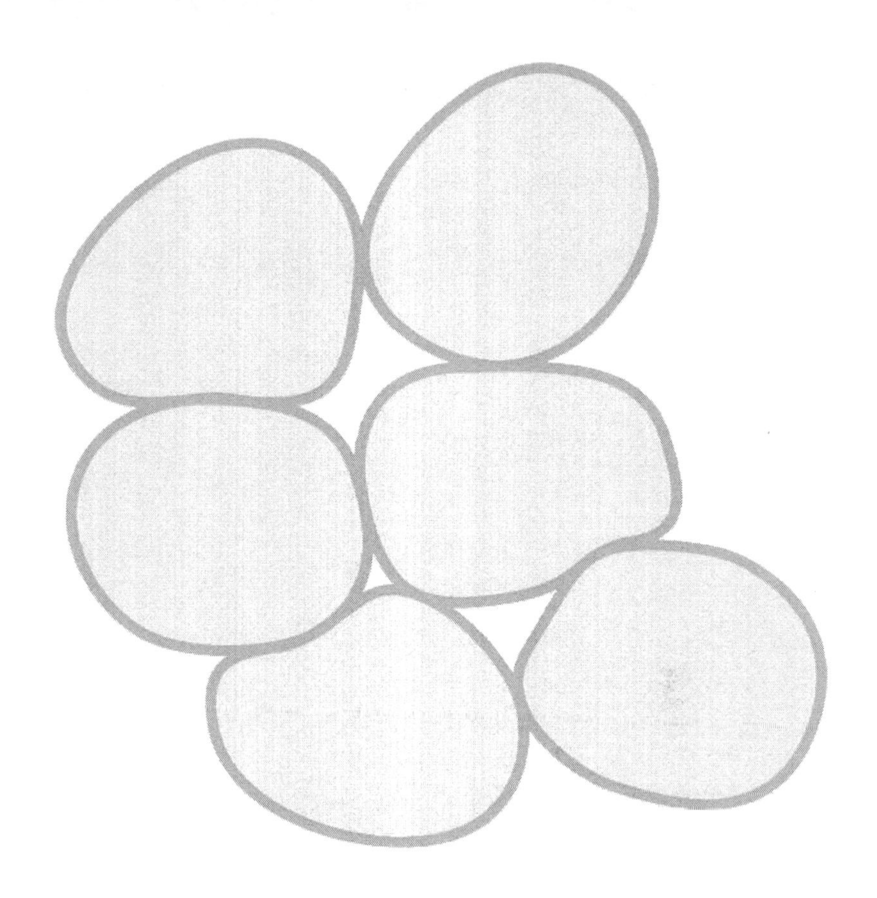

5. How big is a cell?

Cells come in different shapes and sizes. They are not visible to the naked eye. The smallest thing the human eye can see is 0.1 mm long. You need a microscope to see cells.

The smallest type of cell is the "**mycoplasma**" (a type of bacteria) measuring 0.0001 mm in diameter. The longest animal cells are the nerve cells of the giraffe's neck. They are 3 meters or just about 10 feet in length. It is thought that the "laryngeal nerve cells" of the dinosaur "Supersaurus" may have been longer than 28 meters (92 feet).

Human red blood cells measure 0.00076 mm. The largest human cell is the ovum visible by the naked eye.

About 10,000 (ten thousand) average sized human cells can fit on the head of a pin. The head of a pin is around 2 mm.

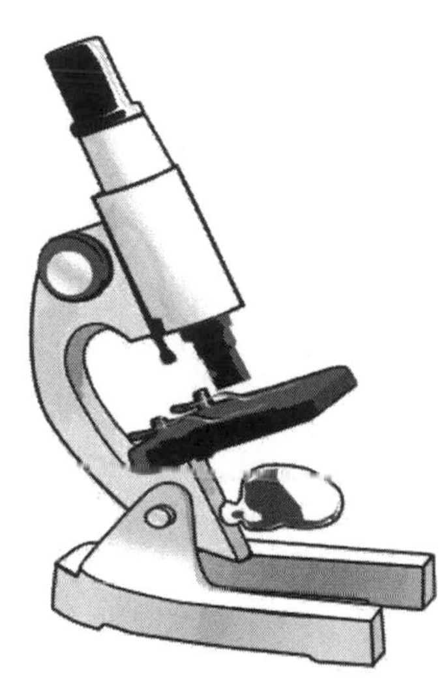

Questions:

1. What is the name of the smallest cell?

2. How long is the longest cell?

3. What instrument do you need to see a cell?

6. How many cells are in the human body?

It is difficult to count all the cells in the body. But we know that there are lots and lots of cells because everything is made of cells. It is thought that there are **100 trillion cells** in the human body.

Cells are born and they die in large numbers all the time. Every cell type has a life span. Taste receptor cells (cells in the taste buds) is replaced every 10 days, about 30 days for skin cells, 15 years for muscle cells, nerve cells (also called **neurons**) remain in the body for whole life.

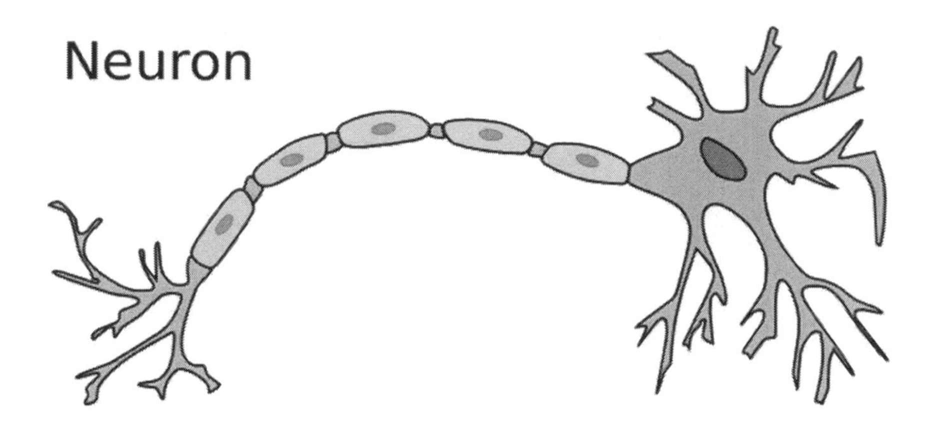

Questions:

1. How many cells are there in the human body?

2. What is the name of cells in the taste buds?

3. What's the other name for nerve cells?

4. How many zeros are there in a trillion?

7. How many different types of cells are there?

In the biological world there are 2 types of cells. They are - cells that have a nucleus and cells that don't have a nucleus.

Cells that have a nucleus are called "**eukaryotes**". The nucleus of eukaryotes contains DNA (De-oxy-ribo-nucleic acid). DNA is like a set of instructions that makes protein telling the cell what jobs need to be done. **Remember**: DNA makes RNA makes protein.

Cells that don't have a nucleus are called "**prokaryotes**" (prokaryotic cells) also known as anuclear cells. They are called prokartotic cells because they do not have nucleus, mitochondria, or any other organelles that are surrounded by membranes.

Most of the prokaryotes are uni-cellular organisms. This means that they are just one-celled organism. The largest single celled organism is the “bubble algae" also called "**sailors eyeballs**” (see picture below).

The adult human body has around 100 trillion cells and around 210 types. Human cells are eukaryotic cells.

You think of your body and there is a cell type for it. Living things are made of different kind of cells that are specialized for different functions. There are cells for your taste buds, there are long nerve cells (also known as neurons in the body), there are cells in the eye called “sensory cells”, there are fat storing cells called “**adipocytes**”, there are function cells for the lung, heart, skin, kidney, muscles and so on, there are blood cells and many more.

All cells are classified according to the roles they play and molecules on their cell surface.

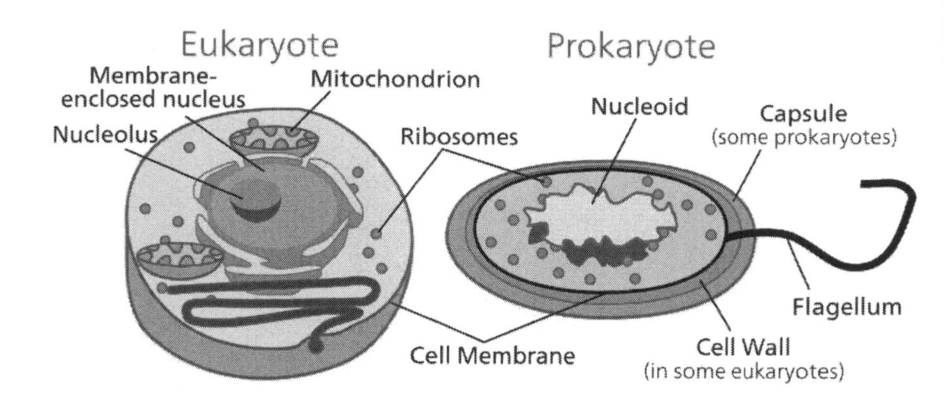

Questions:

1. What are the two types of cells?

2. How many different types of cells are in the human body?

2. What are adipocytes?

8. The animal cell

An “animal cell” is a eukaryotic cell that has various parts called **organelles**. Organelles are surrounded by a membrane. Organelles are structures in the cells that have roles to play by performing a specific function.

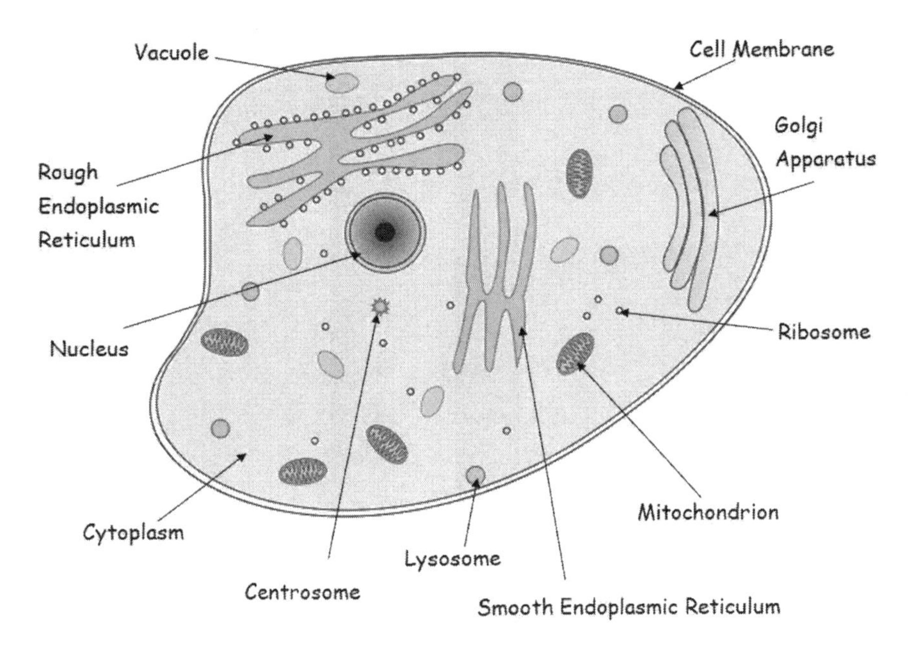

9. Parts and organelles of animal cells

Organelles are membrane-bound (those that have a membrane) particles inside the cell which have specific function.

Cell membrane

Cell membrane is also known as "plasma membrane". It is the outer layer of the cell that controls what gets in and what gets out of the cell. It also helps the cell to maintain its shape and prevents it from falling apart. The cell membrane is made up of a mix of lipids and protein. It is not strictly an organelle.

Nucleus

The Nucleus contains the genetic material DNA (De-oxy-ribo-nucleic acid). The DNA has genes that make proteins which helps controls the activities of the cell. Think of nucleus as the control center of the cell. Inside the nucleus, there is a nucleolus. Can you find the "**nucleolus"** inside the nucleus in the diagram of the cell above? It's the dense dark spot. Sometimes it's called a "sub-organelle". The nucleolus helps to form ribosomes.

Ribosome

Ribosomes are found all around the cell. You will also find them attached to the endoplasmic reticulum. When they are attached to the endoplasmic reticulum, the endoplasmic reticulum is then called the rough endoplasmic reticulum because it looks "rough" under the microscope.

The cells make protein and they make it with the help of ribosomes. Think of ribosomes of as the "protein makers" of the cell. ***Remember****: Ribosomes* *are not considered as organelles as it does not have a membrane.*

Endoplasmic Reticulum

The rough endoplasmic reticulum has ribosomes attached to them while in the smooth endoplasmic reticulum there are no ribosomes.

The rough endoplasmic reticulum is a site where lots of proteins are made all the time such as in a liver cell. It is here where ribosomes make proteins.

The smooth endoplasmic reticulum mainly makes lipids (cholesterol and fat) and carbohydrates for the cells and body to use.

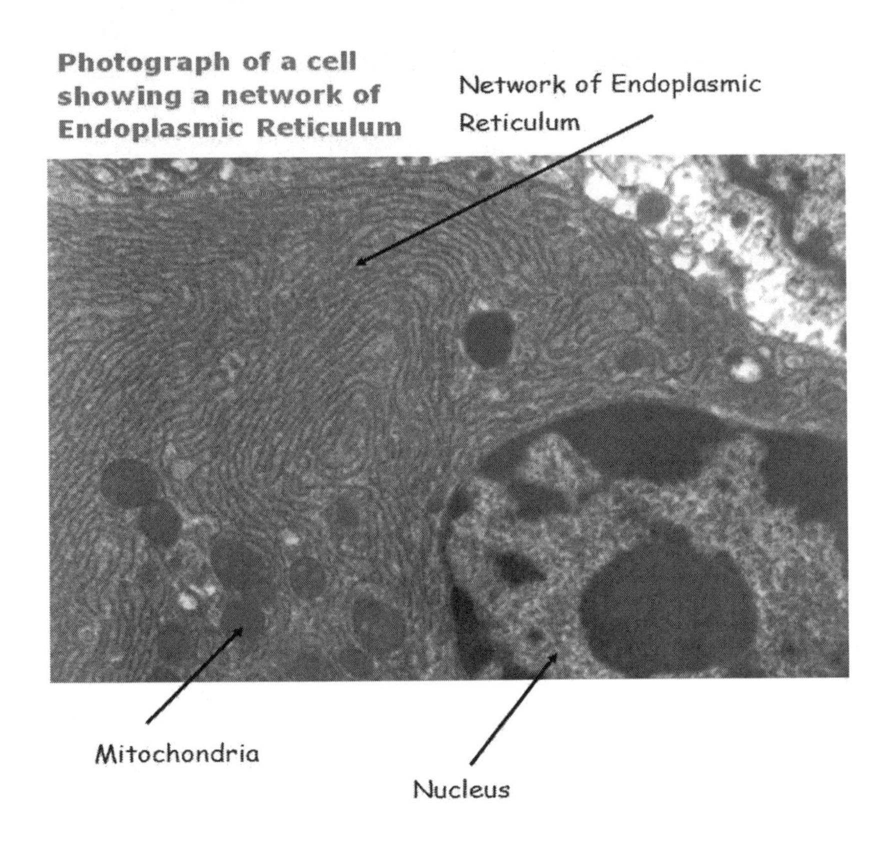

Mitochondria

These are the power-houses of the cell. They make energy that can be used by the cell.

Centrosome

Centrosomes are found in animal cells and are responsible for the cells to divide. It is often located near the nucleus.

Lysosome

These organelles hold enzymes made by the cell. The enzymes help the cell to digest things. The enzymes are made by endoplasmic reticulum and then packed by the "Golgi Apparatus" into round vesicle like structures which we call the "lysosome". They are floating around in the cytoplasm until they are needed by the cell.

Golgi Apparatus

They are also called the Golgi Bodies. They are "sac or folded" like structures responsible for packing proteins and lipids made by the cell so that they can be sent to their destinations.

10. The plant cell

Plant cells are eukaryotic cells. Like animals cells, plant cells too have cytoplasm, nucleus, cell membrane, mitochondria, endoplasmic reticulum and ribosomes. But there are some things that plant cells have that the animal cells don't. They are cell wall, chloroplasts and vacuoles.

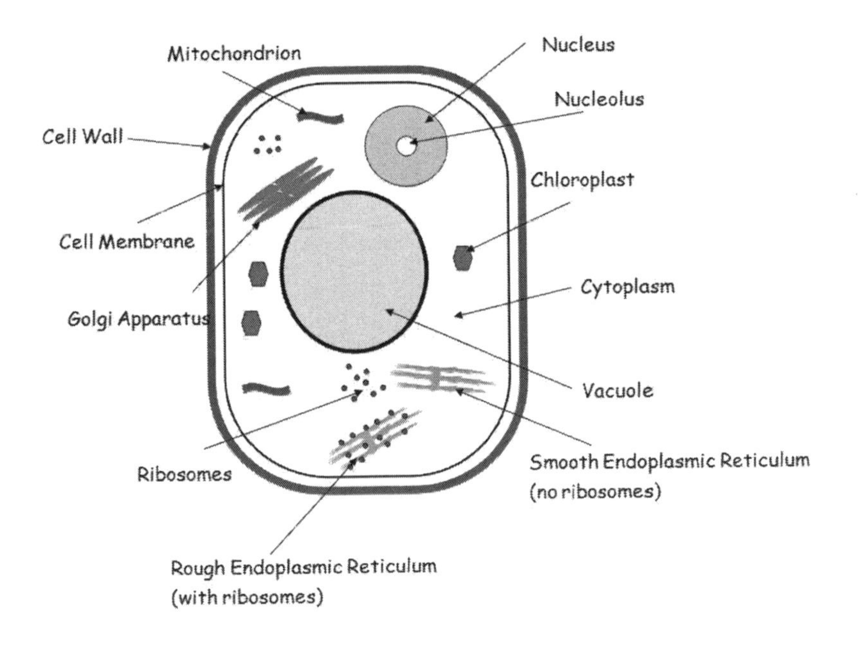

11. The parts and organelles of plant cells

Cell wall

The cell wall is only present in the plant cell. It forms the outermost layer of the plant cell. It helps strengthen the cell and provide protection. It prevents the cell from over expanding when water enters the cell otherwise the cell might burst.

Chloroplast

Chloroplast is a type of plastid. They produce food for the cell. They convert energy from the sunlight to energy storage molecules called ATP (adeno-sine tri-phos-phate) for the plant.

In this process, oxygen is also released in the air which we breathe. This process is called "**photosynthesis**". Chloroplasts have a green pigment that absorbs light called the “**chlorophyll**” important in photosynthesis. Plants are green because of this pigment. The energy produced by the chloroplast is broken down by the mitochondria for the cell to use.

Vacuole

Vacuoles are storage bubbles. They are filled with water along with other things. Some of the other things are enzymes that may be harmful to the cell and waste material generated by the cell. They also help maintain the right acidity of the cell, exporting material away from the cell and providing structure to the cell.

The size of the plant cell is dependent on the vacuoles.

12. Animal cells and plant cells – The differences

Plant cells are generally larger than animal cells.

Chloroplasts, vacuoles and cell wall are not found in animal cells.

Animal cells have lysosomes while plant cells don't.

Plant cells take part in photosynthesis while animal cells don't.

Plant cells don't have centrosomes.

Animal cells can have more than one small vacuole while plant cells have one large vacuole.

Vacuoles are bigger in plant cells than animal cells.

Why animal cells don't have a cell wall?

Animal cells don't have a cell wall. This is because the animal cell doesn't require the support in the same way a plant cell needs. The animal cells have structures that give them extra strength such as exo-skeletons and endo-skeletons (bones of the cells).

Also not having the cell wall allows animal cells to become more specialized to organ cells more easily.

Not having a cell wall makes them more mobile and allows animal cells to move. The plant cell can't move around.

13. What are tissues, organs and organ systems?

Tissues

Cells of the same type form tissues.

Cells of the same type group together in an organized form to make a tissue - both in animals and plants. For example, epithelial cells are packed together to form the lining inside the intestines.

Some other tissues are; muscles, the lining of the lungs, connective tissues (such as blood and bone).

Plants too have tissues. The main ones are the epidermis, the ground tissue, and the vascular tissue.

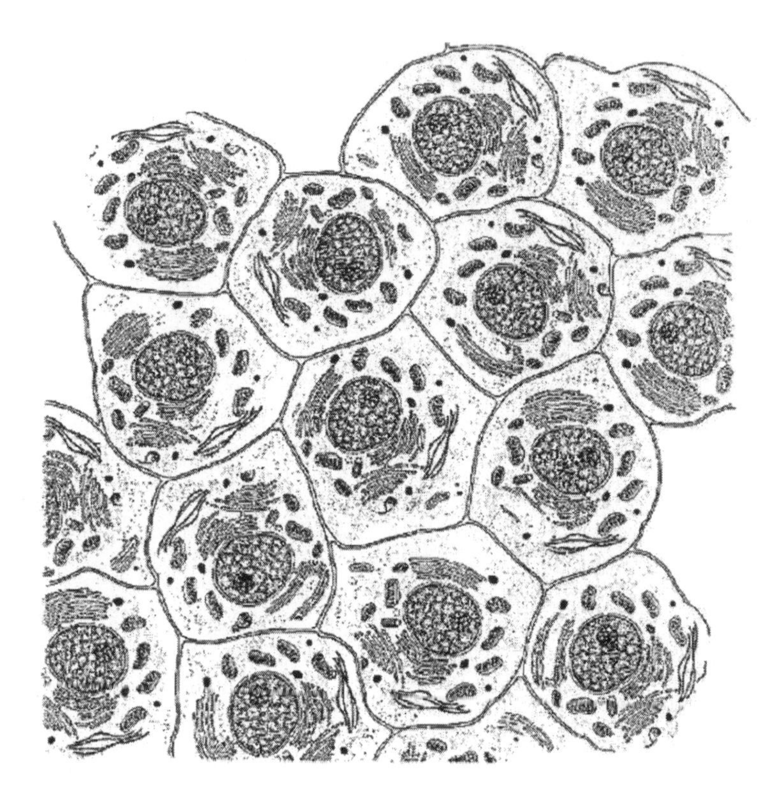

Organs

Different tissues make up the organ that works together for a particular function. Some human and animal organs are heart, liver, lungs and brain. Some plant organs are the roots, flowers, stems and leaves.

Organ Systems

Many different organs form an organ system which work together to do a particular job.

Some organ systems of humans and animals include the “respiratory system” that helps us to breathe and the “digestive system” that helps us to digest food.

Plants too have organs such as “root system”, the “shoot system” and the “reproductive system”.

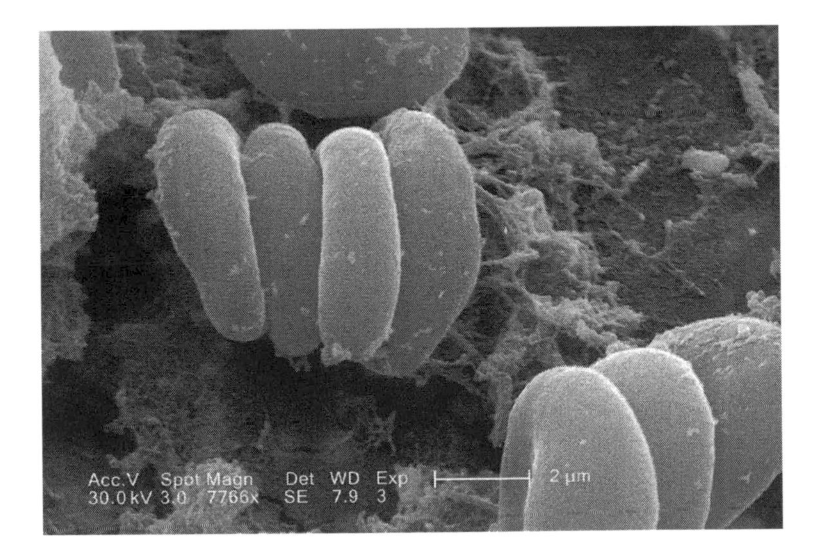

14. Cellular division – Cell cycle

Cell division is a process where one cell divides into two or more cells.

During cell division, the nucleus splits and is passed onto the daughter cells. Cell division is taking place all the time in the human body. Around 2 trillion cell divisions occur in the human body every single day.

There are two types of cells (1) Mitosis and (2) Meiosis.

(1) Mitosis

Mitosis is where the parent cell "**replicates"** to produce two new identical daughter cells. The two cells have the same amount of genetic material - the DNA.

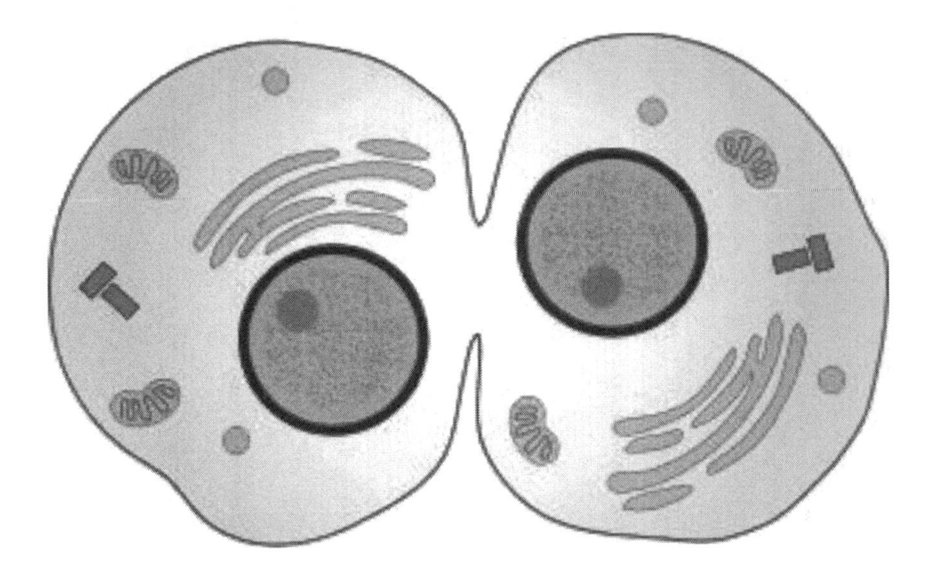

(2) Meiosis

This is the process by which reproductive cells are produced. Meiosis only happens in the testes and the ovaries. Meiosis results in the formation of 4 cells and each cell has half of the genetic material (DNA) of the parent cell. The cells produced are called “haploid” cells.

Remember - in meiosis 1 single cell produces 4 new cells instead of just 2.

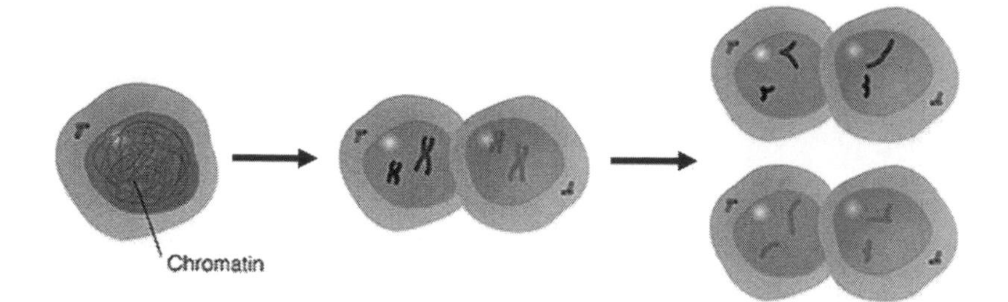

Meiosis: 1 cell divides to form 4 cells.

15. 10 facts about the cell

1. Cells are too small to be seen without a microscope.

2. There are 2 main types of cells – (1) Eukaryotic cells because they have a nucleus and (2) Prokaryotes cells that don't.

3. Eukaryotic cells are bigger than prokaryotic cells because they have a nucleus.

4. Cells contain parts inside them called organelles with different functions to carry out for the cell.

5. Cells of the same type grouped together are called tissues and tissues are grouped to together to form organs. Many organs make up an organ system.

6. Cells can develop features to have specific functions. They become specialized for a particular function in a process called "differentiation".

7. There are more bacterial cells in the human body than our own cells.

8. Cells have a genetic material called DNA. Remember DNA makes RNA which then makes protein. DNA is present in the nucleus.

9. Cells divided to produce new cells. One parent cell will divide into two or more daughter cells. There are two types of cell division: mitosis and meiosis.

10. A cell can self destruct itself in a process called "apoptosis". This process helps prevent cancer.

16. Quiz - What can you remember?

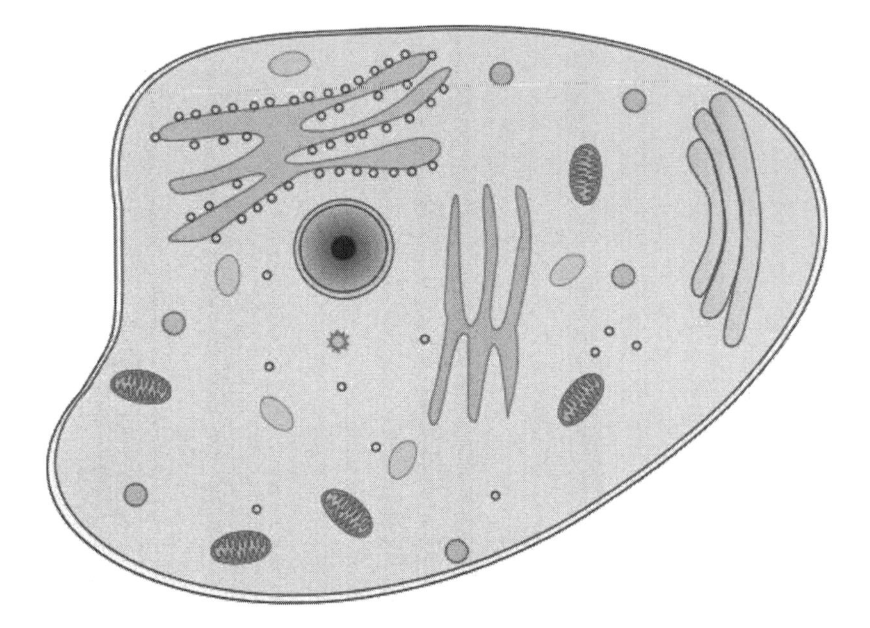

1. Can you name the parts of this animal cell?

2. Which organelle is the "power house" of the cell?

3. How many mitochondria can you see?

4. What is the role of the mitochondria in the cell?

5. How many lysosomes can you see? What does it do?

6. Can you point to the nucleus? What is the function of the nucleus?

7. What does the rough endoplasmic reticulum have that the smooth endoplasmic reticulum doesn't?

8. Can you remember one organelle only found in plant cells?

9. Where is the DNA found in the cell?

10. Which of these is not an organelle? - Mitochondria, nucleus or liver.

11. Which organelle is responsible for making proteins?

12. Which pigment gives green color to plants?

13. Can you draw a cell?

From the author and credits

Thank you for reading this book. If you liked this book then tell your friends about it. Also consider leaving a review for this book. Simply return to Amazon and provide your review.

Photo credits

Mouse cell (in part 1: What is a cell?). Photo taken by Dee Lauzon, Sue Lancelle and Marian Rice of Mount Holyoke College. Public Domain: Cell Image Library.

Red blood cell (in part 11: What are tissues, organs and organ systems) Janice Haney Carr). Photo by Janice Haney Carr. Public Domain: Centers for Disease Control and Prevention.

17289569R00027

Made in the USA
Middletown, DE
16 January 2015